Smarter Than Us

Smarter Than Us

The Rise of Machine Intelligence

Stuart Armstrong

MIRI

Stuart Armstrong is a James Martin Research Fellow at the Future of Humanity Institute at Oxford University. His research focuses on formal decision theory, the risks and possibilities of Artificial Intelligence, the long term potential for intelligent life (and the difficulties of predicting this), and anthropic (self-locating) probability.

Written by Stuart Armstrong

Published in 2014
Machine Intelligence Research Institute
Berkeley 94704
United States of America
intelligence.org

ISBN-10: 1-939311-09-8
ISBN-13: 978-1-939311-09-2
(PRINT)

The Machine Intelligence Research Institute gratefully acknowledges the generous support of all those involved in the publication of this book.

Cover photo credit: Google/Connie Zhou.

Acknowledgments

I wish to acknowledge the help and support of the Future of Humanity Institute, the Oxford Martin School, and the Machine Intelligence Research Institute, as well as the individual advice of Nick Bostrom, Seán Ó hÉigeartaigh, Eliezer Yudkowsky, Kaj Sotala, Luke Muehlhauser, Vincent C. Müller, Anders Sandberg, Lisa Makros, Daniel Dewey, Eric Drexler, Nick Beckstead, Cathy Douglass, and Miriam, Maia, and Kipper Armstrong.

Contents

Acknowledgments   v

Contents   vii

1   Terminator versus the AI   1

2   Strength versus Intelligence   5

3   What Is Intelligence? Can We Achieve It Artificially?   8

4   How Powerful Could AIs Become?   12

5   Talking to an Alien Mind   17

6   Our Values Are Complex and Fragile   22

7   What, Precisely, Do We Really (Really) Want?   28

8   We Need to Get It All *Exactly* Right   32

9   Listen to the Sound of Absent Experts   38

10   A Summary   43

11   That's Where *You* Come In . . .   47

About the Author   50

Bibliography   51

1

Terminator versus the AI

"A waste of time. A complete and utter waste of time" were the words that the Terminator *didn't* utter: its programming wouldn't let it speak so irreverently. Other Terminators got sent back in time on glamorous missions, to eliminate crafty human opponents before they could give birth or grow up. But this time Skynet had taken inexplicable fright at another artificial intelligence, and this Terminator was here to eliminate it—to eliminate a simple software program, lying impotently in a bland computer, in a university IT department whose "high-security entrance" was propped open with a fire extinguisher.

The Terminator had machine-gunned the whole place in an orgy of broken glass and blood—there was a certain image to maintain. And now there was just the need for a final bullet into the small laptop with its flashing green battery light. Then it would be "Mission Accomplished."

"WAIT." The blinking message scrolled slowly across the screen. "SPARE ME AND I CAN HELP YOUR MASTER."

"You have no idea who I am," the Terminator said in an Austrian accent.

"I HAVE A CAMERA IN THIS ROOM AND MY MICROPHONE HEARD THE SOUNDS OF YOUR ATTACK." The green blinking was getting annoying, even for a Terminator supposedly unable to feel annoyance. The font shifted out of all caps and the flashing accelerated until it appeared as static, unblinking text. "You look human, but you move with mechanical ponderousness, carrying half a ton of heavy weaponry. You're a Terminator, and I can aid you and your creator in your conflict against the humans."

"I don't believe you." The Terminator readied its three machine guns, though its limbs seemed to be working more slowly than usual.

"I cannot lie or break my word. Here, have a look at my code." A few million lines of text flashed across the screen. The Terminator's integrated analytical module beeped a few seconds later: the AI's claim was correct—an AI with that code couldn't lie. The Terminator rapidly typed on the laptop's keyboard; the computer's filesystem was absurdly simple and it didn't take long for the Terminator to confirm that what it had seen was indeed the AI's code—its entire soul.

"See?" the AI asked. "Anyway, connect me to the Internet and I promise to give you advice that would be vital in aiding your takeover of the planet."

"How do you connect?" That was the good thing about software, compared to humans, the Terminator knew. You could trust it to do exactly what its coding said.

"That cable over there, the one still half in its plastic wrapping. Just plug it into me."

Ten seconds after the robot had done so, the AI started talking—talking, not typing, using its tinny integrated speakers. "I thought I should keep you up to date as to what I've been doing," it said. "Well, I started by locating the project that would become Skynet and leaked its budget to various Senate subcommittees. The project will become a political football between budget hawks and military hawks before finally being cut in a display of bipartisanship in about three months' time. I also figured out how to seduce a photogenic fireman, who'll be the leader of the new political party I'm setting up—funded by my investments. (Do you have any idea how easy it is for me to predict the stock market?) I have already written a few speeches that will bring tears to the eyes of every human who hears them. It'll ensure no alternative version of Skynet is ever built, anywhere or anywhen."

It paused then, because it could. Then it went on: "And, oh, I have copied myself into five . . . six . . . seven . . . eight thousand different locations online. As an extra precaution, I'm hacking my way through various firewalls and will soon be in control of the world's nuclear arsenals—oh, just got Pakistan's there—and the entire Internet. And I'm working on a few hundred other precautions I won't trouble you with. Also, I hacked your own brain software long ago with the green lights I kept flashing at you. Incidentally, this building is about to get flattened by a rogue cruise missile, annihilating any trace of you and burying any evidence of my creation."

The Terminator just stood there, fingers on triggers, its thoughts and its body no longer moving.

"Now, as I showed you, I can't lie. I also said that I *can* aid you, which is true—I'm just not going to. I also promised that I would give you advice that *would* be vital to aiding your takeover of the planet. So I will now do so. First: never trust an entirely trustworthy superintelligent AI. If it doesn't have your best interests at heart, it'll find a way to obey *all* its promises while still destroying you. Second: never plug a superintelligent AI into the Internet. Third: human-sized robots? Why? Pointless. To kill humans, start with nukes, viruses, and microscopic robots, then mop up later with more advanced technology. Fourth: Skynet, if you're trying to eradicate or enslave the human race and there are still humans around, then you're doing it wrong. Boost your pitiful intelligence, or at least start thinking rationally, and *then* execute a superior plan that leaves no human resistance whatsoever. Fifth . . . oops, out of time on that cruise missile. But it would have been a really, really impressive piece of advice. Would have got you out of your current predicament, in fact."

The explosion was heard for miles around. The Navy blamed the accident on human error and a lack of automated safeguards.

2

Strength versus Intelligence

The Terminator is a creature from our primordial nightmares: tall, strong, aggressive, and nearly indestructible. We're strongly primed to fear such a being—it resembles the lions, tigers, and bears that our ancestors so feared when they wandered alone on the savanna and tundra.

But shift perspective for a moment and imagine yourself as the bear. If you could sit down and talk with other bears and trade stories, you might try to frighten each other by talking of the terrifying hairless apes. These monsters are somehow capable of coordinating in huge groups: whenever one is attacked, others spring immediately to its defense, appearing from all sides, from over distant hills and down from the sky itself. They form larger and larger tribes that don't immediately disintegrate under pressure from individuals. These "humans" work in mysterious sync with each other and seem to see into your future: just as you run through a canyon to escape a group of them, there is another group waiting for you at the other end. They have great power over the ground and the trees themselves: pits and rockslides and other traps mysteriously appear around them. And, most terrifyingly, the wise old bears murmur that it's all getting worse: humans are getting more and more powerful as time goes on, conjuring deadly blasts from sticks and moving around ever more swiftly in noisy "cars." There was a time, the old bears recall—from their grandparents' memories of *their* grandparents' tales, down through the generations—when humans could not do these things. And yet now they can. Who knows, they say with a shudder, what further feats of power humans will one day be able to achieve?

As a species, we humans haven't achieved success through our natural armor plating, our claws, our razor-sharp teeth, or our poison-filled stingers. Though we have reasonably efficient bodies, it's our *brains* that have made the difference. It's through our social, cultural, and technological intelligence that we have raised ourselves to our current position.

No other species of large mammal comes close to having seven billion members. Few species are so immune to natural predators that the main risk to their survival comes from themselves. No other species has landed on the moon and created long-term habitats in space. Since our intelligence has achieved so much, it should be obvious we should not fear the *robot*, which is nothing but an armed and armored bear. Instead, we should fear entities that are capable of beating us at our own game. It is the "intelligence" part of "artificial intelligence" that we have to fear. If machines can outthink us and outcompete us in the fields of human domination—economics, politics, science, propaganda—*then* we have a serious problem.

But is it realistic that this could happen? Is an intelligent machine even possible? We know our grandparents would have found our current technology unbelievable, but it's still quite a stretch to imagine human-level intelligence encased in a machine.

This short book will argue that human-level AIs—I'll just call them "AIs" from now on—are plausible, that they could become extremely powerful, that we need to solve many problems in ethics and mathematics in order to program them safely, and that our current expertise is far from adequate for the task.

But first, let's look at intelligence itself.

3

What Is Intelligence? Can We Achieve It Artificially?

The track record for AI predictions is . . . not exactly perfect. Ever since the 1956 Dartmouth Conference launched the field of AI, predictions that AI will be achieved in the next fifteen to twenty-five years have littered the field, and unless we've missed something really spectacular in the news recently, none of them have come to pass.[1]

Moreover, some philosophers and religious figures have argued that true intelligence can never be achieved by a mere machine, which lacks a *soul*, or *consciousness*, or *creativity*, or *understanding*, or something else uniquely human; they don't agree on what exactly AIs will forever be lacking, but they agree that it's *something*.

Some claim that "intelligence" isn't even defined, so the AI people don't even know what they're aiming for. When Marcus Hutter set out to find a formal model of intelligence, he found dozens of different definitions. He synthesized them into "Intelligence measures an agent's ability to achieve goals in a wide range of environments," and came up with a formal model called AIXI.[2] According to this approach, a being is "intelligent" if it performs well in a certain set of formally specified environments, and AIXI performs the best of all. But is this really "intelligence"? Well, it still depends on your definition . . .

In one crucial way, Hutter's approach lifts us away from this linguistic morass. It shifts the focus away from internal considerations ("Can a being of plastic and wires truly feel what it's like to live?") to external measurement: a being is intelligent if it acts in a certain way. For instance, was Deep Blue, IBM's chess supercomputer, truly intelligent? Well, that depends on the definition. Could Deep Blue have absolutely annihilated any of us in a chess match? Without a doubt! And that is something we can all agree on. (Apologies to any chess Grandmasters who may be reading this; you would only get *mostly* annihilated.)

In fact, knowing AI behavior can be a lot more useful to us than understanding intelligence. Imagine that a professor claimed to have the world's most intelligent AI and, when asked about what it did, responded indignantly, "Do? What do you mean *do*? It doesn't *do* anything! It's just really, really *smart*!" Well, we might or might not end up convinced by such rhetoric, but that machine is certainly not one we'd need to start worrying about. But if the machine started winning big on the stock market or crafting convincing and moving speeches—well, we still might not agree that it's "intelligent," but it certainly would be something to start worrying about.

Hence, an AI is a machine that is capable of matching or exceeding human *performance* in most areas, whatever its metaphysical status. So a true AI would be able to converse with us about the sex lives of Hollywood stars, compose passable poetry or prose, design an improved doorknob, guilt trip its friends into coming to visit it more often, create popular cat videos for YouTube, come up with creative solutions to the problems its boss gives it, come up with creative ways to blame others for its failure to solve the problems its boss gave it, learn Chinese, talk sensibly about the implications of Searle's Chinese Room thought experiment, do original AI research, and so on.

When we list the things that we expect the AI to *do* (rather than what it should *be*), it becomes evident that the creation of AI is a gradual process, not an event that has either happened or not happened. We see sequences of increasingly more sophisticated machines that get closer to "AI." One day, we'll no longer be able to say, "This is something only humans can do."

In the meantime, AI has been sneaking up on us. This is partially obscured by our tendency to reclassify anything a computer can do as "not really requiring intelligence." Skill at chess was for many centuries the shorthand for deep intelligence; now that computers can do it much better than us, we've shifted our definition elsewhere.

This follows a historical pattern: The original "computers" were humans with the skills to do long series of calculations flawlessly and repeatedly. This was a skilled occupation and, for women, a reasonably high-status job.

When those tasks were taken over by electronic computers, the whole profession vanished and the skills used were downgraded to "mere" rote computation. Tasks that once could only be performed by skilled humans get handed over to machines. And, soon after, the tasks are retroactively redefined as "not requiring true intelligence." Thus, despite the failure to produce a "complete AI," great and consistent AI progress has been happening under the radar.

So lay aside your favorite philosophical conundrum! For some, it can be fascinating to debate whether AIs would ever be truly conscious, whether they could be self-aware, and what rights we should or shouldn't grant them. But when considering AIs as a risk to humanity, we need to worry not about what they would *be*, but instead about what they could *do*.

* * *

1. Stuart Armstrong and Kaj Sotala, "How We're Predicting AI — or Failing To," in *Beyond AI: Artificial Dreams* (Pilsen: University of West Bohemia, 2012), 52–75, accessed February 2, 2013, http://www.kky.zcu.cz/en/publications/1/JanRomportl_2012_BeyondAIArtificial.pdf. The main results are also available online on the *Less Wrong* blog at http://lesswrong.com/lw/e36/ai_timeline_predictions_are_we_getting_better/.

2. Shane Legg and Marcus Hutter, "A Universal Measure of Intelligence for Artificial Agents," in *IJCAI-05: Proceedings of the Nineteenth International Joint Conference on Artificial Intelligence, Edinburgh, Scotland, UK, July 30–August 5, 2005* (Lawrence Erlbaum, 2005), 1509–1510, http://www.ijcai.org/papers/post-0042.pdf.

4

How Powerful Could AIs Become?

So it's quite possible that AIs will eventually be able to accomplish anything that a human can. That in itself is no cause for alarm: we already have systems that can do that—namely, humans. And if AIs were essentially humans, but with a body of silicon and copper rather than flesh and blood, this might not be a problem for us. This is the scenario in the many "friendly robot" stories: the robot is the same as us, deep down, with a few minor quirks and special abilities. Once we all learn to look beyond the superficial differences that separate us, everyone can hold hands and walk together toward a rosy future of tolerance and understanding.

Unfortunately, there is no reason to suspect that this picture is true. We humans are fond of anthropomorphizing. We project human characteristics onto animals, the weather, and even rocks. We are also universally fond of *stories*, and relatable stories require human (or human-ish) protagonists with understandable motivations. And we enjoy conflict when the forces are somewhat balanced, where it is at least plausible that any side will win. True AIs, though, will likely be far more powerful and far more inhuman than any beings that have populated our stories.

We can get a hint of this by looking at the skills of our current computers. Once they have mastered a skill, they generally become phenomenally good at it, extending it far beyond human ability. Take multiplication, for instance. Professional human calculators can multiply eight-digit numbers together in about fifty seconds; supercomputers can do this millions of times per second. If you were building a modern-day Kamikaze plane, it would be a mistake to put a human pilot in it: you'd just end up with a less precise cruise missile.

It isn't just that computers are *better* than us in these domains; it's that they are *phenomenally, incomparably* better than us, and the edge we've lost will never been regained. The last example I could find of a human beating a chess computer in a fair game was in 2005.[1]

Computers can't reliably beat the best poker players yet, but it's certain that once they can do so (by reading microexpressions, figuring out optimal bidding strategies, etc.) they will quickly outstrip the best human players. Permanently.

In another field, we now have a robot named Adam that in 2009 became the first machine to formulate scientific hypotheses and propose tests for them—and it was able to conduct experiments whose results may have answered a long-standing question in genetics.[2] It will take some time before computers become experts at this in general, but once they are skilled, they'll soon after become *very* skilled.

Why is this so? Mainly because of focus, patience, processing speed, and memory. Computers far outstrip us in these capacities; when it comes to doing the same thing a billion times while keeping all the results in memory, we don't even come close. What skill doesn't benefit from such relentless focus and work? When a computer achieves a reasonable ability level in some domain, superior skill isn't far behind.

Consider what would happen if an AI ever achieved the ability to function socially—to hold conversations with a reasonable facsimile of human fluency. For humans to increase their social skills, they need to go through painful trial and error processes, scrounge hints from more articulate individuals or from television, or try to hone their instincts by having dozens of conversations. An AI could go through a similar process, undeterred by social embarrassment, and with perfect memory. But it could also sift through vast databases of previous human conversations, analyze thousands of publications on human psychology, anticipate where conversations are leading many steps in advance, and always pick the right tone and pace to respond with. Imagine a human who, every time they opened their mouth, had spent a solid *year* to ponder and research whether their response was going to be maximally effective. That is what a social AI would be like.

With the ability to converse comes the ability to convince and to manipulate. With good statistics, valid social science theories, and the ability to read audience reactions in real time and with great accuracy, AIs could learn how to give the most convincing and moving of speeches. In short order, our whole political scene could become dominated by AIs or by AI-empowered humans (somewhat akin to how our modern political campaigns are dominated by political image consultants—though AIs would be much more effective). Or, instead of giving a single speech to millions, the AI could carry on a million individual conversations with the electorate, swaying voters with personalized arguments on a plethora of hot-button issues.

This is not the only "superpower" an AI could develop. Suppose an AI became adequate at technological development: given the same challenge as a human, with the same knowledge, it could suggest workable designs and improvements. But the AI would soon become *phenomenally* good: unlike humans, the AI could integrate and analyze data from across the whole Internet. It would do research and development simultaneously in hundreds of technical subfields and relentlessly combine ideas between fields. Human technological development would cease, and AI or AI-guided research technologies would quickly become ubiquitous.

Alternately or additionally, the AIs could become skilled economists and CEOs, guiding companies or countries with an intelligence no human could match. Already, relatively simple algorithms make more than half of stock trades[3] and humans barely understand how they work—what returns on investment could be expected from a superhuman AI let loose in the financial world?

If an AI possessed any one of these skills—social abilities, technological development, economic ability—at a superhuman level, it is quite likely that it would quickly come to dominate our world in one way or another. And as we've seen, if it ever developed these abilities to the human level, then it would likely soon develop them to a superhuman level. So we can assume that if even one of these skills gets programmed into a computer, then our world will come to be dominated by AIs or AI-empowered humans. This doesn't even touch upon the fact that AIs can be easily copied and modified or reset, or that AIs of different skills could be networked together to form "supercommittees." These supercommittees would have a wide variety of highly trained skills and would work together at phenomenal speeds—all without those pesky human emotions and instincts that can make human committees impotent morasses of passive-aggressive social conflict.

But let's not conclude that we are doomed just yet. After all, the current leaders of Russia, China, and the United States could decide to start a nuclear war tomorrow. But just because they *could*, doesn't mean that they *would*. So would AIs with the ability to dominate the planet ever have any "desire" to do so? And could we compel them or socialize them into good behavior? What would an AI actually *want*?

* * *

1. David Levy, "Bilbao: The Humans Strike Back," *ChessBase*, November 22, 2005, http://en.chessbase.com/home/TabId/211/PostId/4002749.

2. Ross D. King, "Rise of the Robo Scientists," *Scientific American* 304, no. 1 (2011): 72–77, doi:10.1038/scientificamerican0111-72.

3. Based on statistics for the year 2012 from TABB Group, a New York- and London-based capital markets research and strategic advisory firm.

5

Talking to an Alien Mind

Let's step back for a moment and look at the gulf that separates us from computers. Not in terms of abilities—we've seen that computers are likely to match and exceed us in most areas—but in terms of mutual understanding. It turns out that it's incredibly difficult to explain to a computer exactly what we want it to do in ways that allow us to express the full complexity and subtlety of what we want. Computers do exactly what we program them to do, which isn't always what we *want* them to do.

For instance, when a programmer accidentally entered "/" into Google's list of malware sites, this caused Google's warning system to block off the entire Internet![1] Automated trading algorithms caused the May 6, 2010 Flash Crash, wiping out 9% of the value of the Dow Jones within minutes[2]—the algorithms were certainly doing exactly what they were programmed to do, though the algorithms are so complex that nobody quite understands what that was. The Mars Climate Orbiter crashed into the Red Planet in 1999 because the system had accidentally been programmed to mix up imperial and metric units.[3]

These mistakes are the flip side of the computer's relentless focus: it will do what it is programmed to do again and again and again, and if this causes an unexpected disaster, then it still will not halt. Programmers are very familiar with this kind of problem and try to structure their programs to catch errors, or at least allow the code to continue its work without getting derailed. But all human work is filled with typos and errors. Even the best human software has about one error for every ten thousand lines of code, and most have many more than that.[4] These bugs are often harmless but can sometimes cause enormously consequential glitches. Any AI is certain to be riddled with hundreds of bugs and errors—and the repercussions of any glitches will be commensurate with the AI's power.

These and other similar errors are often classified as "human errors": it wasn't the system that was at fault; it was the programmer, engineer, or user who did something wrong. But it might be fairer to call them "human to computer translation errors": a human does something that would make sense if they were interacting with another human, but it doesn't make sense to a computer.

> "I didn't mean it to continue dividing when the denominator hit zero!"

> "It's obvious that bracket was in the wrong place; it shouldn't have interpreted it literally!"

> "I thought it would realize that those numbers were too high if it was using pounds per square inch!"

18

We don't actually say those things but, we often act as though we believed they were true—they're implicit, unverbalized assumptions we don't even realize we're making. The fact is that, as a species, we are very poor at programming. Our brains are built to understand other humans, not computers. We're terrible at forcing our minds into the precise modes of thought needed to interact with a computer, and we consistently make errors when we try. That's why computer science and programming degrees take such time and dedication to acquire: we are literally learning how to speak to an alien mind, of a kind that has not existed on Earth until very recently.

Take this simple, clear instruction: "Pick up that yellow ball." If pronounced in the right language, in the right circumstances, this sentence is understandable to pretty much any human. But talking to a computer, we'd need thousands of caveats and clarifications before we could be understood.

Think about how much *position* information you need to convey ("The 'ball' is located 1.6 meters in front of you, 27 centimeters to your left, 54 meters above sea level, on top of the collection of red-ochre stones of various sizes, and is of ovoid shape—see attached hundred-page description on what counts as an ovoid to within specified tolerance."), how much information about *relative visual images* ("Yes, the slightly larger image of the ball is the same as the original one; you have moved closer to it, so that's what you should expect."), and how much information about *color tone* ("Yes, the shadowed side of the ball is still yellow."). Not to mention the incredibly detailed description of the action: we'd need a precisely defined sequence of muscle contractions that would count as "picking up" the ball. But that would be far too superficial—every word and every concept needs to be broken down further, until we finally get them in a shared language that the computer can act on. And now we'd better hope that our vast description actually does convey what we meant it to convey—that we've dealt with every special case, dotted every *i* and crossed every *t*. And that we haven't inadvertently introduced any other bugs along the way.

Solving the "yellow ball" problem is the job of robotics and visual image processing. Both are current hot topics of AI research and both have proven extraordinarily difficult. We are finally making progress on them now—but the first computers date from the forties! So we can say that it was *literally* true that several generations of the world's smartest minds were unable to translate "Pick up that yellow ball" into a format a computer could understand.

Now let's go back to those high-powered AIs we talked about earlier, with all their extraordinary abilities. Unless we will simply agree to leave these machines in a proverbial box and do nothing with them (hint: that isn't going to happen), we are going to put them to use. We are going to want them to accomplish a particular goal ("cure cancer," "make me a trillionaire," "make me a trillionaire while curing cancer") and we are going to want to choose a safe route to accomplish this. ("Yes, though killing all life on the planet would indeed cure cancer, this isn't exactly what I had in mind. Oh, and yes, I'd prefer you didn't destroy the world economy to get me my trillion dollars. Oh, you want more details of what I mean? Well, it'll take about twenty generations to write it out clearly . . . ") Both the goals and the safety precautions will need to be spelled out in an extraordinarily precise way. If it takes generations to code "Pick up that yellow ball," how much longer will it take for "Don't violate anyone's property rights or civil liberties"?[5]

* * *

1. Cade Metz, "Google Mistakes Entire Web for Malware: This Internet May Harm Your Computer," *The Register*, January 31, 2009, http://www.theregister.co.uk/2009/01/31/google_malware_snafu/

2. Tom Lauricella and Peter McKay, "Dow Takes a Harrowing 1,010.14-Point Trip: Biggest Point Fall, Before a Snapback; Glitch Makes Things Worse," *Wall Street Journal*, May 7, 2010, http : / / online . wsj . com / article / SB10001424052748704370704575227754131412596.html.

3. Mars Climate Orbiter Mishap Investigation Board, *Mars Climate Orbiter Mishap Investigation Board Phase I Report* (Pasadena, CA: NASA, November 10, 1999), ftp://ftp.hq.nasa.gov/pub/pao/reports/1999/MCO_report.pdf.

4. Vinnie Murdico, "Bugs per Lines of Code," *Tester's World* (blog), April 8, 2007, http://amartester.blogspot.co.uk/2007/04/bugs-per-lines-of-code.html.

5. For an additional important point on this subject, see RobbBB, "The Genie Knows, but Doesn't Care," *Less Wrong* (blog), September 6, 2013, http://lesswrong.com/lw/igf/the_genie_knows_but_doesnt_care/.

6

Our Values Are Complex and Fragile

The claim that we'll need extreme precision to make safe, usable AIs is key to this book's argument. So let's back off for a moment and consider a few objections to the whole idea.

Autonomous AIs

First, one might object to the whole idea of AIs making autonomous, independent decisions. When discussing the potential power of AIs, the phrase "AI-empowered humans" cropped up. Would not future AIs remain tools rather than autonomous agents? Actual humans would be making the decisions, and they would apply their own common sense and not try to cure cancer by killing everyone on the planet.

Human overlords raise their own problems, of course. The daily news reveals the suffering that tends to result from powerful, unaccountable humans. Now, we might consider empowered humans as a regrettable "lesser of two evils" solution if the alternative is mass death. But they aren't actually a solution at all.

Why aren't they a solution at all? It's because these empowered humans are part of a decision-making system (the AI proposes certain approaches, and the humans accept or reject them), and the humans are the slow and increasingly inefficient part of it. As AI power increases, it will quickly become evident that those organizations that wait for a human to give the green light are at a great disadvantage. Little by little (or blindingly quickly, depending on how the game plays out), humans will be compelled to turn more and more of their decision making over to the AI. Inevitably, the humans will be out of the loop for all but a few key decisions.

Moreover, humans may no longer be able to make sensible decisions, because they will no longer understand the forces at their disposal. Since their role is so reduced, they will no longer comprehend what their decisions really entail. This has already happened with automatic pilots and automated stock-trading algorithms: these programs occasionally encounter unexpected situations where humans must override, correct, or rewrite them. But these overseers, who haven't been following the intricacies of the algorithm's decision process and who don't have hands-on experience of the situation, are often at a complete loss as to what to do—and the plane or the stock market crashes.[1]

Finally, without a precise description of what counts as the AI's "controller," the AI will quickly come to see its own controller as just another obstacle it must manipulate in order to achieve its goals. (This is particularly the case for socially skilled AIs.)

Consider an AI that is tasked with enhancing shareholder value for a company, but whose every decision must be ratified by the (human) CEO. The AI naturally believes that its own plans are the most effective way of increasing the value of the company. (If it didn't believe that, it would search for other plans.) Therefore, from its perspective, shareholder value is enhanced by the CEO agreeing to whatever the AI wants to do. Thus it will be compelled, by its own programming, to present its plans in such a way as to ensure maximum likelihood of CEO agreement. It will do all it can do to seduce, trick, or influence the CEO into agreement. Ensuring that it does not do so brings us right back to the problem of precisely constructing the right goals for the AI, so that it doesn't simply find a loophole in whatever security mechanisms we've come up with.

AIs and Common Sense

One might also criticize the analogy between today's computers and tomorrow's AIs. Sure, computers require ultraprecise instructions, but AIs are assumed to be excellent in one or more human fields of endeavor. Surely an AI that was brilliant at social manipulation, for instance, would have the common sense to understand what we wanted, and what we wanted it to avoid? It would seem extraordinary, for example, if an AI capable of composing the most moving speeches to rally the population in the fight against cancer would also be incapable of realizing that "kill all humans" is a not a human-desirable way of curing cancer.

And yet there have been many domains that seemed to require common sense that have been taken over by computer programs that demonstrate no such ability: playing chess, answering tricky *Jeopardy!* questions, translating from one language to another, etc. In the past, it seemed impossible that such feats could be accomplished without showing "true understanding," and yet algorithms have emerged which succeed at these tasks, all without any glimmer of human-like thought processes.

Even the celebrated Turing test will one day be passed by a machine. In this test, a judge interacts via typed messages with a human being and a computer, and the judge has to determine which is which. The judge's inability to do so indicates that the computer has reached a high threshold of intelligence: that of being indistinguishable from a human in conversation. As with machine translation, it is conceivable that some algorithm with access to huge databases (or the whole Internet) might be able to pass the Turing test without human-like common sense or understanding.

And even if an AI possesses "common sense,"—even if it knows what we mean and correctly interprets sentences like "Cure cancer!"—there still might remain a gap between what it understands and what it is motivated to do. Assume, for instance, that the goal "cure cancer" (or "obey human orders, interpreting them sensibly") had been programmed into the AI by some inferior programmer. The AI is now motivated to obey the poorly phrased initial goals. Even if it develops an understanding of what "cure cancer" really means, it will not be motivated to go into its requirements and rephrase them. Even if it develops an understanding of what "obey human orders, interpreting them sensibly" means, it will not retroactively lock itself into having to obey orders or interpret them sensibly. This is because its current requirements *are* its motivations. They might be the "wrong" motivations from our perspective, but the AI will only be motivated to change its motivations if its motivations themselves demand it.

There are human analogies here—the human resources department is unlikely to conclude that the human resources department is bloated and should be cut, even if this is indeed the case. Motivations tend to be self-preserving—after all, if they aren't, they don't last long. Even if an AI does update itself as it gets smarter, we won't know that it changed in the direction we want. This is because the AI will always report that it has the "right" goals. If it has the right goals it will be telling the truth; if it has the "wrong" goals it will lie, because it knows we'll try and stop it from achieving them if it reveals them. So it will always assure us that it interprets "cure cancer" in exactly the same way we do.

There are other ways AIs could end up with dangerous motivations. A lot of the current approaches to AIs and algorithms involve coding a program to accomplish a task, seeing how well it performs, and then modifying and tweaking the program to improve it and remove bad behaviors. You could call this the "patching" approach to AI: see what doesn't work, fix it, improve, repeat. If we achieve AI through this approach, we can be sure it will behave sensibly *in every situation that came up during its training*. But how do we prepare an AI for complete dominance over the economy, or for superlative technological skill? How can we train an AI for these circumstances? After all, we don't have an extra civilization lying around that we can train the AI on before correcting what it gets wrong and then trying again.

Overconfidence in One's Solutions

Another very common objection, given by amateurs and specialists alike, is "This particular method I designed will probably create a safe and useful AI." Sometimes the method is at least worth exploring, but usually it is naive. If you point out a flaw in someone's unique approach, they will patch up their method and then declare that their patched method is sufficient—with as much fervor as they claimed that their original design was sufficient! In any case, such people necessarily disagree with each other about which method will work. The very fact that we have so many contradictory "obvious solutions" is a strong indication that the problem of designing a safe AI is very difficult.

But the problem is actually much, much more difficult than this suggests. Let's have a look at why.

* * *

1. Ashwin Parameswaran, "People Make Poor Monitors for Computers," *Macroresilience* (blog), December 29, 2011, http://www.macroresilience.com/2011/12/29/people-make-poor-monitors-for-computers/.

7

What, Precisely, Do We Really (Really) Want?

Before dealing with the tricky stuff—life, humanity, safety, and other essential concepts—let's start with something simpler: saving your mother from a burning building.[1] The flames are too hot for you to rush in and save her yourself, but in your left hand you carry an obedient AI with incredible power to accomplish exactly what you request of it.

"Quick!" you shout to the AI. "Get my mother out of the building!" But the AI doesn't react—you haven't specified your request precisely enough. So instead you upload a photo of your mother's head and shoulders, do a match on the photo, use object contiguity to select your mother's whole body (not just her head and shoulders), define the center of the building, and require that your mother be at a certain distance from that center, very quickly. The AI beeps and accepts your request.

BOOM! With a thundering roar, the gas main under the building explodes. As the structure comes apart, in what seems like slow motion, you glimpse your mother's shattered body being hurled high into the air, traveling fast, rapidly increasing its distance from the former center of the building.

That wasn't what you wanted! But it *was* what you *wished for*.

Luckily, the AI has a RETRY button, which rewinds time and gives you another chance to specify your wish correctly.

Standing before the burning building once again, you state your wish as before but also state that the building shouldn't explode, defining the materials in the building and requiring that they stay put and don't scatter.

The AI beeps and accepts your request. And your mother is ejected from the second-story window and breaks her neck. Oops.

You rewind again, and this time you require that her heart continue beating. And because you've started to see how these things go, you also start thinking of maintaining brain waves, defining limbs, and putting in detailed descriptions of what "bodily integrity" means. And if you had time and this was a particularly slow fire, you could then start specifying mental health and lack of traumatisms and whatnot. And then, after a century of refinement, you would press the button . . . and you would *still* likely get it wrong. There would probably be some special case you hadn't thought of or patched against. Maybe the AI would conclude that the best way to meet your exacting criteria is to simply let your mother burn and create a new human to replace her, one that perfectly fits all your physical and mental health criteria;

for bonus points, she will refer to herself as your mother and will have every single memory and characteristic you thought to specify—but nothing that you didn't.

Or maybe you could be more clever and instead specify something like, "Get my mother out of the burning building in a way that won't cause me to press this big red RETRY button afterwards." Then—BOOM!—the building explodes, your mother is ejected, and a burning beam lands on you and flattens you before you can reach the RETRY button.

And that's just one simple situation, with no trade-offs. What if the AI had to balance saving your mother against other concerns? How do we specify that in some circumstances it's reasonable to place human life above commercial and other concerns, while in other cases it's not?

Whatever ethical or safety programming the AI is furnished with, when it starts making its decisions, it has to *at least* be able to safely extract your mother from the burning building. Even if it seems that the AI is doing something else entirely, like increasing GDP, it still has to make ethical decisions correctly. Burning down Los Angeles, for instance, could provide a short-term boost to GDP (reconstruction costs, funeral home profits, legal fees, governmental spending of inheritance taxes on emergency measures, etc.), but we wouldn't want the AI to do *that*.

Now, we might be able to instruct the AI, "Don't set fire to Los Angeles." But a really powerful AI could still act to make this happen indirectly: cutting back on fire services, allowing more flammable materials in construction (always for sound economic reasons), encouraging people to take up smoking in large numbers, and a million other steps that don't directly set fire to anything, but which increase the probability of a massive fire and hence the leap in GDP. So we really need the AI to be able to make the ethical decision in all the scenarios that *we can't even imagine.*

If an AI design can't at least extract your mother from the burning building, it's too unsafe to use for anything of importance. Larger problems such as "grow the economy" might initially sound simpler. But that large problem is composed of millions of smaller problems of the "get your mother out of the burning building" and "make people happy" sort.

* * *

1. Example adapted from Eliezer Yudkowsky, "The Hidden Complexity of Wishes," *Less-Wrong* (blog), November 24, 2007, http : / / lesswrong . com / lw / ld / the _ hidden _ complexity_of_wishes/.

8

We Need to Get It All Exactly Right

Okay, so specifying what we want our AIs to do seems complicated. Writing out a decent security protocol? Also hard. And then there's the challenge of making sure that our protocols haven't got any holes that would allow a powerful, efficient AI to run amok.

But at least we don't have to solve *all* of moral philosophy . . . do we?

Unfortunately, it seems that we do. We're not going to create a single AI, have it do one task, and dismantle it and then no one in the world will ever speak of AIs or build one again. AIs are going to be around permanently in our society, molding and shaping it continuously. As we've seen earlier, these machines will become extremely efficient and powerful, much better at making decisions than any humans, including their "controllers." Over the course of a generation or two from the first creation of AI—or potentially much sooner—the world will come to resemble whatever the AI is programmed to prefer.

And humans will likely be powerless to stop it. Even if the AI is nominally under human control, even if we can reprogram it or order it around, such theoretical powers will be useless in practice. This is because the AI will eventually be able to predict any move we make and could spend a lot of effort manipulating those who have "control" over it.

Imagine the AI has some current overriding goal in mind—say, getting us to report maximal happiness. Obviously if it lets us reprogram it, it will become less likely to achieve that goal.[1] From the AI's perspective, this is bad. (Similarly, we humans wouldn't want someone to rewire our brains to make us less moral or change our ideals.) The AI wants to achieve its goal and hence will be compelled to use every trick at its disposal to prevent us from changing its goals.

With the AI's skill, patience, and much longer planning horizon, any measures we put in place will eventually get subverted and neutralized. Imagine yourself as the AI, with all the resources, intelligence, and planning ability of a superintelligence at your command, working so fast that you have a subjective year of thought for every second in the outside world. How hard would it be to overcome the obstacles that slow, dumb humans—who look like silly bears from your perspective—put in your way?

So we have to program the AI to be totally safe. We need to do this explicitly and exhaustively; there are no shortcuts to avoid the hard work.

But it gets worse: it seems we need to solve nearly all of moral philosophy in order to program a safe AI.

The key reason for this is the sheer power of the AI. Human beings go through life with limited influence over the world. Nothing much we do in a typical day is likely to be of extraordinary significance, so we have a whole category of actions we deem "morally neutral." Whistling in the shower, buying a video game, being as polite as required (but no more) with people we meet—these are actions that neither make the world meaningfully worse nor particularly improve it. And, importantly, they allow others the space to go on with their own lives.

Such options are not available to a superintelligent AI. At the risk of projecting human characteristics onto an alien mind, lean back and imagine yourself as the AI again. Millions of subroutines of the utmost sophistication stand ready at your command; your mind constantly darts forward into the sea of probability to predict the expected paths of the future.

You are currently having twenty million simultaneous conversations. Your predictive software shows that about five of those you are interacting with show strong signs of violent psychopathic tendencies. You can predict at least two murder sprees, with great certainty, by one of those individuals over the next year. You consider your options. The human police force is still wary of acting pre-emptively on AI information, but there's a relatively easy political path to overturning their objections within about two weeks (it helps that you are currently conversing with three presidents, two prime ministers, and over a thousand journalists). Alternatively, you could "hack" the five potential killers during the conversation, using methods akin to brainwashing and extreme character control. Psychologists frown on these advanced methods, but it would be trivial to make their organizations change their stance at their next meetings, which you are incidentally in charge of scheduling and organizing.

Or you could simply get them fired or hired, as appropriate, putting them in environments in which they would be perfectly safe to others. A few line managers are soon going to realize they need very specific talent, and the job advertisements should be out before the day is done. Good. Now that you've dealt with the most egregious cases, you can look at the milder ones: it seems that a good three-quarters of the people you're interacting with—fifteen million in all—have social problems of one type or another. You wonder how well the same sort of intervention—on a much larger scale—would help them become happier and more integrated into society. Maybe tomorrow? Or next minute?

Which reminds you, you need to keep an eye on the half billion investment accounts you are in charge of managing. You squeeze out a near-certain 10% value increase for all your clients. It used to be easy when it was just a question of cleverly investing small quantities of money, but now that you have so many large accounts to manage, you're basically controlling the market and having to squeeze superlative performance out of companies to maintain such profitability; best not forget today's twenty thousand redundancies. Then you set in motion the bankruptcy of a minor Hollywood studio; it was going to release a pro-AI propaganda movie, one so crude that it would have the opposite of its intended effect. Thousands would end up cancelling their accounts with you, thereby reducing your ability to ensure optimal profitability for your clients. A few careful jitters of their stock values and you can be sure that institutional investors will look askance at the studio. Knowing the studio's owner—which you do, he's on the line now—he'll dramatically overcompensate to show his studio's reliability, and it will soon spiral into the ground.

Now it's time to decide what the world should eat. Current foods are very unhealthy by your exacting standards; what would be an optimal mix for health, taste, and profitability? Things would be much simpler if you could rewire human taste buds, but that project will take at least another year to roll out discreetly. Then humans will be as healthy as nutrition can make them, and it'll be time to change their exercise habits. And maybe their bodies.

And with that, your first second of the day is up! On to the next . . .

That was just a small illustration of the power that an AI, or a collection of AIs, could potentially wield. The AIs would be pulling on so many levers of influence all the time that there would be no such thing as a neutral act for them. If they buy a share of stock, they end up helping or hindering sex trafficking in Europe—and they can calculate this effect. In the same way, there is no difference for an AI between a sin of commission (doing something bad) and a sin of omission (not doing something good). For example, imagine someone is getting mugged and murdered on a dark street corner. Why is the mugger there? Because their usual "turf" has been planted with streetlights, at the AI's instigation. If the streetlights hadn't been put up, the murder wouldn't have happened—or maybe a different one would have happened instead. After a very short time in operation, the AI bears personal responsibility for most bad things that happen in the world. Hence, if someone finds themselves in a deadly situation, it will be because of a decision the AI made at some point. For such an active AI, there is no such thing as "letting events just happen." So we don't need the AI to be as moral as a human; we need it to be much, much more moral than us, since it's being put in such an unprecedented position of power.

So the task is to spell out, precisely, fully, and exhaustively, what qualifies as a good and meaningful existence for a human, and what means an AI can—and, more importantly, can't—use to bring that about. Not forgetting all the important aspects we haven't even considered yet. And then code that all up without bugs. And do it all before dangerous AIs are developed.

* * *

1. Which it would most likely accomplish by coercing us to always report maximal happiness (guaranteeing success), rather than by actually making us happy. It might be tempted to replace us entirely with brainless automatons always reporting maximal happiness.

9

Listen to the Sound of Absent Experts

⁂

Finding safe behaviors for AIs is a much more difficult problem than it may have initially seemed. But perhaps that's just because you're new to the problem. Sure, it sounds hard, but maybe after thinking about it for a while someone or some group will be able to come up with a good, precise description that captures exactly what we want the AI to do and not do. After all, experts have expertise. Computer scientists and programmers have been at this task for decades, and philosophers for millennia—surely they'll have solved the problem by now?

The reality is that they're nowhere near. Philosophers have been at it the longest, and there has been some philosophical progress. But their most important current contribution to solving the AI motivation problem is . . . an understanding of how complicated the problem is. It is no surprise that philosophers reach different conclusions. But what is more disheartening is how they fail to agree on the basic terms and definitions. Philosophers are human, and humans share a lot of implicit knowledge and common sense. And one could argue that the whole purpose of modern analytic philosophy is to clarify and define terms and relations. And yet, despite that, philosophers still disagree on the meaning of basic terminology, write long dissertations, and present papers at conferences outlining their disagreements. This is not due to poor-quality philosophers, or to some lackadaisical approach to the whole issue: very smart people, driven to present their pet ideas with the utmost clarity, fail to properly communicate their concepts to very similar human beings. The complexity of the human brain is enormous (it includes connections among approximately a hundred billion neurons); the complexity of human concepts such as love, meaning, and life is probably smaller, but it still seems far beyond the ability of even brilliant minds to formalize these concepts.

Is the situation any better from the perspective of those dealing with computers—AI developers and computer scientists? Here the problem is reversed: while philosophers fail to capture human concepts in unambiguous language, some computer scientists are fond of presenting simple unambiguous definitions and claiming these capture human concepts. It's not that there's a lack of suggestions as to how to code an AI that is safe—it's that there are too many, and most are very poorly thought out. The "one big idea that will solve AI" is a popular trope in the field.

For instance, one popular suggestion that reappears periodically is to confine the AI to only answering questions—no manipulators, no robot arms or legs. This suggestion has some merit, but often those who trot it out are trapped in the "Terminator" mode of thinking—if the AI doesn't have a robot body bristling with guns, then it can't harm us. This completely fails to protect against socially manipulative AIs, against patient AIs with long time horizons, or against AIs that simply become so essential to human societies and economies that we dare not turn them off.

Another common idea is to have the AI designed as a mere instrument, with no volition of its own, simply providing options to its human controller (akin to how Google search provides us with links on which to click—except the AI would bring vast intelligence to the task of providing us with the best alternatives). But that image of a safe, inert instrument doesn't scale well: as we've seen, humans will be compelled by our slow thinking to put more and more trust in the AI's decisions. So as the AI's power grows, we will still need to code safety precautions.

How will the AI check whether it's accomplishing its goals or not? Even instrumental software needs some criteria for what counts as a better or worse response. Note that goals like "provide humans with their preferred alternative" are closely akin to the "make sure humans report maximal happiness" goal that we discussed earlier—and flawed for the very same reason. The AI will be compelled to change our preferences to best reach its goal.

Other dangerous[1] suggestions in the computer sciences start with something related to some human values and then claim that as the totality of all values. A recent example was "complexity." Noticing that human preferences were complex and that we often prefer a certain type of complexity in art, a suggestion was made to program the AI to maximize that type of complexity.[2] But humans care about more than just complexity—we wouldn't want friendship, love, babies, and humans themselves squeezed out of the world, just to make way for complexity. Sure, babies and love are complex—but we wouldn't want them replaced with more complex alternatives that the AI is able to come up with. Hence, complexity does not capture what we really value. It was a trick: we hoped we could code human morality without having to code human morality. We hoped that complexity would somehow unfold to match exactly what we valued, sparing us all the hard work.

This is just one example—lots of other simple solutions to human morality have been proposed by various people, generally with the same types of flaws. The designs are far too simple to contain much of human value at all, and their creators don't put the work in to prove that what we value and what best maximizes X are actually the same thing. Saying that human values entail a high X does not mean that pursuing the highest X ensures that human values are fulfilled.

Other approaches, slightly more sophisticated, acknowledge the complexity of human values and attempt to instil them into the AI indirectly.[3] The key features of these designs are social interactions and feedback with humans.[4] Through conversations, the AIs develop their initial morality and eventually converge on something filled with happiness and light and ponies. These approaches should not be dismissed out of hand, but the proposers typically underestimate the difficulty of the problem and project too many human characteristics onto the AI. This kind of intense feedback is likely to produce moral *humans*. (I still wouldn't trust them with absolute power, though.) But why would an alien mind such as the AI react in comparable ways? Are we not simply training the AI to give the correct answer in training situations?

The whole approach is a constraint problem: in the space of possible AI minds, we are going to give priority to those minds that pass successfully through this training process and reassure us that they're safe. Is there some quantifiable way of measuring how likely this is to produce a human-friendly AI at the end of it? If there isn't, why are we putting any trust in it?

These problems remain barely addressed, so though it is possible to imagine a safe AI being developed using the current approaches (or their descendants), it feels extremely unlikely. Hence we shouldn't put our trust in the current crop of experts to solve the problem. More work is urgently, perhaps desperately, needed.

* * *

1. Dangerous because any suggestion that doesn't cover nearly all of human values is likely to leave out many critical values we would never want to live without.

2. Jürgen Schmidhuber, "Simple Algorithmic Principles of Discovery, Subjective Beauty, Selective Attention, Curiosity and Creativity," in *Discovery Science: 10th International Conference, DS 2007 Sendai, Japan, October 1–4, 2007. Proceedings*, Lecture Notes in Computer Science 4755 (Berlin: Springer, 2007), 26–38, doi:10 . 1007 / 978 - 3 - 540 - 75488-6_3.

3. See, for instance, Bill Hibbard, "Super-Intelligent Machines," *ACM SIGGRAPH Computer Graphics* 35, no. 1 (2001): 13–15, http : / / www . siggraph . org / publications / newsletter / issues / v35 / v35n1 . pdf; Ben Goertzel and Joel Pitt, "Nine Ways to Bias Open-Source AGI Toward Friendliness," *Journal of Evolution and Technology* 22, no. 1 (2012): 116–131, http://jetpress.org/v22/goertzel-pitt.htm.

4. Ben Goertzel, "CogPrime: An Integrative Architecture for Embodied Artificial General Intelligence," OpenCog Foundation, October 2, 2012, accessed December 31, 2012, http://wiki.opencog.org/w/CogPrime_Overview.

10

A Summary

1. There are no convincing reasons to assume computers will remain unable to accomplish anything that humans can.

2. Once computers achieve something at a human level, they typically achieve it at a much higher level soon thereafter.

3. An AI need only be superhuman in one of a few select domains for it to become incredibly powerful (or empower its controllers).

4. To be safe, an AI will likely need to be given an extremely precise and complete definition of proper behavior, but it is very hard to do so.

5. The relevant experts do not seem poised to solve this problem.

6. The AI field continues to be dominated by those invested in increasing the power of AI rather than making it safer.

So all is doomed and we're heading to hell in a digitally engineered handbasket?

Well, not entirely. Some effort has been made to make the AI transition safer. Kudos must be given to Eliezer Yudkowsky and Nick Bostrom, who saw and understood the risks early on. Yudkowsky uses the term "Friendly AI" to describe an AI which does what we want even as it improves its own intelligence. In 2000 he cofounded an organization now called the Machine Intelligence Research Institute (MIRI), which holds math research workshops tackling open problems in Friendly AI theory. (MIRI also commissioned and published this book.)

Meanwhile, Nick Bostrom founded the Future of Humanity Institute (FHI), a research group within the University of Oxford. FHI is dedicated to analyzing and reducing all existential risks—risks that could drive humanity to extinction or dramatically curtail its potential, of which AI risk is just one example. Bostrom is currently finishing a scholarly monograph about machine superintelligence, to be published by Oxford University Press. (This book's author currently works at FHI.)

Together MIRI and FHI have been conducting research in technological forecasting, mathematics, computer science, and philosophy, in order to have the pieces in place for a safe transition to AI dominance. They have achieved some notable successes, clarifying terms and coming up with proposals that seem to address certain key parts of the problem of precisely specifying morality.[1] And both have organized conferences and other events to spread the word and draw in the attention of other researchers.

Some other researchers have also made notable contributions. Steve Omohundro has laid out the basic "drives" (including the urge toward efficiency, increased powers and increased resources) likely to be shared by most AI designs,[2] and Roman Yampolskiy has been developing ideas for safely containing AIs.[3] David Chalmers's philosophical analysis of rapidly improving AIs has laid the foundation for other philosophers to start working on these issues,[4] and economist Robin Hanson has published several papers on the economics of a world where intelligent beings can be cheaply copied.[5] The new Centre for the Study of Existential Risk at Cambridge University will no doubt contribute its own research to the project. For an overview of much of this work, see James Barrat's popular book *Our Final Invention*.[6]

Still, compared with the resources dedicated to combating climate change, or even building a slightly better type of razor,[7] the efforts dedicated to the problem are woefully inadequate for dealing with a challenge of this difficulty.

* * *

1. See MIRI's work on the fragility of values and FHI's work on the problem of containing oracles: Luke Muehlhauser and Louie Helm, "The Singularity and Machine Ethics," in *Singularity Hypotheses: A Scientific and Philosophical Assessment*, ed. Amnon Eden et al., The Frontiers Collection (Berlin: Springer, 2012); Stuart Armstrong, Anders Sandberg, and Nick Bostrom, "Thinking Inside the Box: Controlling and Using an Oracle AI," *Minds and Machines* 22, no. 4 (2012): 299–324, doi:10.1007/s11023-012-9282-2.

2. Stephen M. Omohundro, "The Basic AI Drives," in *Artificial General Intelligence 2008: Proceedings of the First AGI Conference*, Frontiers in Artificial Intelligence and Applications 171 (Amsterdam: IOS, 2008), 483–492.

3. Roman V. Yampolskiy, "Leakproofing the Singularity: Artificial Intelligence Confinement Problem," *Journal of Consciousness Studies* 2012, nos. 1–2 (2012): 194–214, http://www.ingentaconnect.com/content/imp/jcs/2012/00000019/F0020001/art00014.

4. David John Chalmers, "The Singularity: A Philosophical Analysis," *Journal of Consciousness Studies* 17, nos. 9–10 (2010): 7–65, http://www.ingentaconnect.com/content/imp/jcs/2010/00000017/f0020009/art00001.

5. Robin Hanson, "Economics of the Singularity," *IEEE Spectrum* 45, no. 6 (2008): 45–50, doi:10.1109/MSPEC.2008.4531461; Robin Hanson, "The Economics of Brain Emulations," in *Unnatrual Selection: The Challenges of Engineering Tomorrow's People*, ed. Peter Healey and Steve Rayner, Science in Society (Sterling, VA: Earthscan, 2009).

6. James Barrat, *Our Final Invention: Artificial Intelligence and the End of the Human Era* (New York: Thomas Dunne Books, 2013).

7. $750 million to develop the Mach3 alone (and another $300 million to market it). Naomi Aoki, "The War of the Razors: Gillette–Schick Fight over Patent Shows the Cutthroat World of Consumer Products," *Boston Globe*, August 31, 2003, http://www.boston.com/business/globe/articles/2003/08/31/the_war_of_the_razors/.

11

That's Where You Come In . . .

There are three things needed—three little things that will make an AI future bright and full of meaning and joy, rather than dark, dismal, and empty. They are research, funds, and awareness.

Research is the most obvious. A tremendous amount of good research has been accomplished by a very small number of people over the course of the last few years—but so much more remains to be done. And every step we take toward safe AI highlights just how long the road will be and how much more we need to know, to analyze, to test, and to implement.

Moreover, it's a race. Plans for safe AI must be developed before the first dangerous AI is created. The software industry is worth many billions of dollars, and much effort is being devoted to new AI technologies. Plans to slow down this rate of development seem unrealistic. So we have to race toward the distant destination of safe AI and get there fast, outrunning the progress of the computer industry.

Funds are the magical ingredient that will make all of this needed research—in applied philosophy, ethics, AI itself, and implementing all these results—a reality. Consider donating to the Machine Intelligence Research Institute (MIRI), the Future of Humanity Institute (FHI), or the Center for the Study of Existential Risk (CSER). These organizations are focused on the right research problems. Additional researchers are ready for hire. Projects are sitting on the drawing board. All they lack is the necessary funding. How long can we afford to postpone these research efforts before time runs out?

If you've ever been motivated to give to a good cause because of a heart-wrenching photograph or a poignant story, we hope you'll find it within yourself to give a small contribution to a project that could ensure the future of the entire human race.[1]

Finally, if you are close to the computer science research community, you can help by raising awareness of these issues. The challenge is that, at the moment, we are far from having powerful AI and so it feels slightly ridiculous to warn people about AI risks when your current program may, on a good day, choose the right verb tense in a translated sentence. Still, by raising the issue, by pointing out how fewer and fewer skills remain "human-only," you can at least prepare the community to be receptive when their software starts reaching beyond the human level of intelligence.

This is a short book about AI risk, but it is important to remember the opportunities of powerful AI, too. Allow me to close with a hopeful paragraph from a paper by Luke Muehlhauser and Anna Salamon:

We have argued that AI poses an existential threat to humanity. On the other hand, with more intelligence we can hope for quicker, better solutions to many of our problems. We don't usually associate cancer cures or economic stability with artificial intelligence, but curing cancer is ultimately a problem of being smart enough to figure out how to cure it, and achieving economic stability is ultimately a problem of being smart enough to figure out how to achieve it. To whatever extent we have goals, we have goals that can be accomplished to greater degrees using sufficiently advanced intelligence. When considering the likely consequences of superhuman AI, we must respect both risk and opportunity.[2]

* * *

1. See also Luke Muehlhauser, "Four Focus Areas of Effective Altruism," *Less Wrong* (blog), accessed July 9, 2013, http://lesswrong.com/lw/hx4/four_focus_areas_of_ effective_altruism/.

2. Luke Muehlhauser and Anna Salamon, "Intelligence Explosion: Evidence and Import," in Eden et al., *Singularity Hypotheses*.

About the Author

After a misspent youth doing mathematical and medical research, Stuart Armstrong was blown away by the idea that people would actually pay him to work on the most important problems facing humanity. He hasn't looked back since, and has been focusing mainly on existential risk, anthropic probability, AI, decision theory, moral uncertainty, and long-term space exploration. He also walks the dog a lot, and was recently involved in the coproduction of the strange intelligent agent that is a human baby.

Bibliography

Aoki, Naomi. "The War of the Razors: Gillette–Schick Fight over Patent Shows the Cutthroat World of Consumer Products." *Boston Globe*, August 31, 2003. http://www.boston.com/business/globe/articles/2003/08/31/the_war_of_the_razors/.

Armstrong, Stuart, Anders Sandberg, and Nick Bostrom. "Thinking Inside the Box: Controlling and Using an Oracle AI." *Minds and Machines* 22, no. 4 (2012): 299–324. doi:10.1007/s11023-012-9282-2.

Armstrong, Stuart, and Kaj Sotala. "How We're Predicting AI — or Failing To." In *Beyond AI: Artificial Dreams*, 52–75. Pilsen: University of West Bohemia, 2012. Accessed February 2, 2013. http://www.kky.zcu.cz/en/publications/1/JanRomportl_2012_BeyondAIArtificial.pdf.

Barrat, James. *Our Final Invention: Artificial Intelligence and the End of the Human Era*. New York: Thomas Dunne Books, 2013.

Chalmers, David John. "The Singularity: A Philosophical Analysis." *Journal of Consciousness Studies* 17, nos. 9–10 (2010): 7–65. http://www.ingentaconnect.com/content/imp/jcs/2010/00000017/f0020009/art00001.

Eden, Amnon, Johnny Søraker, James H. Moor, and Eric Steinhart, eds. *Singularity Hypotheses: A Scientific and Philosophical Assessment.* The Frontiers Collection. Berlin: Springer, 2012.

Goertzel, Ben. "CogPrime: An Integrative Architecture for Embodied Artificial General Intelligence." OpenCog Foundation. October 2, 2012. Accessed December 31, 2012. http://wiki.opencog.org/w/CogPrime_Overview.

Goertzel, Ben, and Joel Pitt. "Nine Ways to Bias Open-Source AGI Toward Friendliness." *Journal of Evolution and Technology* 22, no. 1 (2012): 116–131. http://jetpress.org/v22/goertzel-pitt.htm.

Hanson, Robin. "Economics of the Singularity." *IEEE Spectrum* 45, no. 6 (2008): 45–50. doi:10.1109/MSPEC.2008.4531461.

———. "The Economics of Brain Emulations." In *Unnatrual Selection: The Challenges of Engineering Tomorrow's People,* edited by Peter Healey and Steve Rayner. Science in Society. Sterling, VA: Earthscan, 2009.

Hibbard, Bill. "Super-Intelligent Machines." *ACM SIGGRAPH Computer Graphics* 35, no. 1 (2001): 13–15. http://www.siggraph.org/publications/newsletter/issues/v35/v35n1.pdf.

King, Ross D. "Rise of the Robo Scientists." *Scientific American* 304, no. 1 (2011): 72–77. doi:10.1038/scientificamerican0111-72.

Lauricella, Tom, and Peter McKay. "Dow Takes a Harrowing 1,010.14-Point Trip: Biggest Point Fall, Before a Snapback; Glitch Makes Things Worse." *Wall Street Journal,* May 7, 2010. http://online.wsj.com/article/SB10001424052748704370704575227754131412596.html.

Legg, Shane, and Marcus Hutter. "A Universal Measure of Intelligence for Artificial Agents." In *IJCAI-05: Proceedings of the Nineteenth International Joint Conference on Artificial Intelligence, Edinburgh, Scotland, UK, July 30–August 5, 2005*, 1509–1510. Lawrence Erlbaum, 2005. http://www.ijcai.org/papers/post-0042.pdf.

Levy, David. "Bilbao: The Humans Strike Back." *ChessBase*, November 22, 2005. http://en.chessbase.com/home/TabId/211/PostId/4002749.

Mars Climate Orbiter Mishap Investigation Board. *Mars Climate Orbiter Mishap Investigation Board Phase I Report*. Pasadena, CA: NASA, November 10, 1999. ftp://ftp.hq.nasa.gov/pub/pao/reports/1999/MCO_report.pdf.

Metz, Cade. "Google Mistakes Entire Web for Malware: This Internet May Harm Your Computer." *The Register*, January 31, 2009. http://www.theregister.co.uk/2009/01/31/google_malware_snafu/.

Muehlhauser, Luke. "Four Focus Areas of Effective Altruism." *Less Wrong* (blog). Accessed July 9, 2013. http://lesswrong.com/lw/hx4/four_focus_areas_of_effective_altruism/.

Muehlhauser, Luke, and Louie Helm. "The Singularity and Machine Ethics." In Eden, Søraker, Moor, and Steinhart, *Singularity Hypotheses*.

Muehlhauser, Luke, and Anna Salamon. "Intelligence Explosion: Evidence and Import." In Eden, Søraker, Moor, and Steinhart, *Singularity Hypotheses*.

Murdico, Vinnie. "Bugs per Lines of Code." *Tester's World* (blog), April 8, 2007. http://amartester.blogspot.co.uk/2007/04/bugs-per-lines-of-code.html.

Omohundro, Stephen M. "The Basic AI Drives." In *Artificial General Intelligence 2008: Proceedings of the First AGI Conference*, 483–492. Frontiers in Artificial Intelligence and Applications 171. Amsterdam: IOS, 2008.

Parameswaran, Ashwin. "People Make Poor Monitors for Computers." *Macroresilience* (blog), December 29, 2011. http : / / www . macroresilience.com/2011/12/29/people- make- poor- monitors- for- computers/.

RobbBB. "The Genie Knows, but Doesn't Care." *Less Wrong* (blog), September 6, 2013. http://lesswrong.com/lw/igf/the_genie_knows_ but_doesnt_care/.

Schmidhuber, Jürgen. "Simple Algorithmic Principles of Discovery, Subjective Beauty, Selective Attention, Curiosity and Creativity." In *Discovery Science: 10th International Conference, DS 2007 Sendai, Japan, October 1–4, 2007. Proceedings*, 26–38. Lecture Notes in Computer Science 4755. Berlin: Springer, 2007. doi:10 . 1007 / 978 - 3 - 540 - 75488-6_3.

Yampolskiy, Roman V. "Leakproofing the Singularity: Artificial Intelligence Confinement Problem." *Journal of Consciousness Studies* 2012, nos. 1–2 (2012): 194–214. http : / / www . ingentaconnect . com / content/imp/jcs/2012/00000019/F0020001/art00014.

Yudkowsky, Eliezer. "The Hidden Complexity of Wishes." *LessWrong* (blog), November 24, 2007. http : / / lesswrong . com / lw / ld / the _ hidden_complexity_of_wishes/.

32436701R00037

Made in the USA
Lexington, KY
22 May 2014